Balancing Chemical Equations

© Learning Hub Publishing. All Rights Reserved. No Part of this publication must be reproduced, distributed or transmitted in any form or by any means without prior written permission of the publisher.

Table of Contents

Basics of Balancing Equations 1

Balance Equation in 3 Terms 5

Balance Equation in 4 Terms 12

Balance Equation in 5 Terms 21

Chemical Equation Word Problems 25

Balance Chemical Equations

Basics of Balancing a Chemical Equation

In Chemistry we should realize when a chemical equation needs to be balanced and if it is not then we go ahead and do it.

In a balanced cheicl equation there are same number of atoms of each type on either side of the reactant.

This means that if you have 2 atoms of Nitrogen on the left side then you must have 2 atoms of Nitrogen on the right and if you have 4 atoms of Oxygen on left side of the reactant then you should have 4 on the right side as well.

One thing you should know is what are coefficients and the atoms and how do we know what they are as they play an important role while balancing the equations.

Balance Chemical Equations

So the number in front of the equation represents the moles of a compound whereas the number in the subscript of the atom show number of atoms in single molecule.

One must also remember the below law,

<u>Law of Conservation of mass:</u>

<u>According to the law of conservation of mass, the mass of products that are derived from a chemical equation should equal the mass of the reactants.</u>

To calculate the number of atoms you should multiply the coefficient and the subscript.

A Balanced equation must be reduced to its lowest whole number coefficient thus go ahead and divide it by 2 or 3 until you get so.

Now Lets check out a few Examples,

Balance Chemical Equations

Balance the Below Chemical Equation

$Na_2CO_3 + HCl \longrightarrow NaCl + CO_2 + H_2O$

We go ahead and see that Sodium is not balanced and is in single state thus we balance it.

$Na_2CO_3 + HCl \longrightarrow 2NaCl + CO_2 + H_2O$

Now we see that there are two cl on the right but only one on the left so we balance that out.

$Na_2CO_3 + 2HCl \longrightarrow 2NaCl + CO_2 + H_2O$

This balances our entire equation and we can see that the even the oxygen and hydrogen atoms are well balanced.

We take a look at another example,

Balance Chemical Equations

Balance the Below Chemical Equation

$$C_2H_6 + O_2 \rightarrow H_2O + CO_2$$

We first start by changing the coefficients of C and H.

$$C_2H_6 + O_2 \rightarrow 3H_2O + 2CO_2$$

This results in 7 O atoms in the product side this we use a fractional coefficient which is $\frac{7}{2}$.

$$C_2H_6 + \frac{7}{2}O_2 \rightarrow 3H_2O + 2CO_2$$

We multiply both the sides with 2 to get a balanced equation

$$2C_2H_6 + 7O_2 \rightarrow 6H_2O + 4CO_2$$

Balance Chemical Equation in 3 Terms

1) $Fe + O_2 \rightarrow Fe_2O_3$

2) $C + S_8 \rightarrow CS_2$

3) $N_2 + H_2 \rightarrow NH_3$

4) $P + O_2 \rightarrow P_2O_5$

Balance Chemical Equation in 3 Terms

5) $H_2 + I_2 \rightarrow HI$

6) $H_2 + Cl_2 \rightarrow HCl$

7) $Na + O_2 \rightarrow NaO_2$

8) $AgO_2 \rightarrow Ag + O_2$

Balance Chemical Equation in 3 Terms

9) $P_4 + Cl_2 \rightarrow PCl_3$

10) $HgO \rightarrow Hg + O_2$

11) $N_2 + O_2 \rightarrow N_2O_5$

12) $S_8 + O_2 \rightarrow SO_3$

Balance Chemical Equation in 3 Terms

13) $Fe + O_2 \rightarrow Fe_2O_3$

14) $Al + S_8 \rightarrow Al_2S_3$

15) $H_2O + O_2 \rightarrow H_2O_2$

16) $PtCl_4 \rightarrow Pt + Cl_2$

Balance Chemical Equation in 3 Terms

17) $Cs + N_2 \rightarrow Cs_3N$

18) $S_8 + F_2 \rightarrow SF_6$

19) $Al + O_2 \rightarrow Al_2O_3$

20) $P_4O_{10} + H_2O \rightarrow H_3PO_4$

Answers

1) $4Fe + 3O_2 \rightarrow 2FeO_3$

2) $4C + 1S_8 \rightarrow 4CS_2$

3) $N_2 + 3H_2 \rightarrow 2NH_3$

4) $4P + 5O_2 \rightarrow 2P_2O_5$

5) $H_2 + I_2 \rightarrow 2HI$

6) $H_2 + Cl_2 \rightarrow 2HCl$

7) $4Na + 1O_2 \rightarrow 2NaO_2$

8) $2AgO_2 \rightarrow 4Ag + 1O_2$

9) $P_4 + 6Cl_2 \rightarrow 4PCl_3$

10) $2HgO \rightarrow 2Hg + O_2$

Answers

11) $N_2 + O_2 \rightarrow N_2O_5$

12) $1S_8 + 12O_2 \rightarrow 8SO_3$

13) $4Fe + 3O_2 \rightarrow 2Fe_2O_3$

14) $16Al + 3S_8 \rightarrow 8Al_2S_3$

15) $2H_2O + 1O_2 \rightarrow 2H_2O_2$

16) $PtCl_4 \rightarrow Pt + 2Cl_2$

17) $6Cs + 1N_2 \rightarrow 2Cs_3N$

18) $S_8 + 24F_2 \rightarrow 8SF_6$

19) $4Al + 3O_2 \rightarrow 2Al_2O_3$

20) $P_4O_{10} + 6H_2O \rightarrow 4H_3PO_4$

Balance Chemical Equation in 4 Terms

1) Ag + H_2S → Ag_2S + H_2

2) K + B_2O_3 → KO_2 + B

3) HCOOH + O_2 → 2 CO_2 + H_2O

4) NaCl + F_2 → NaF + Cl_2

Balance Chemical Equation in 4 Terms

5) $CH_4 + O_2 \rightarrow CO_2 + H_2O$

6) $Co + H_2O \rightarrow Co_2O_3 + H_2$

7) $Na + NaNO_3 \rightarrow Na_2O + N_2$

8) $ZnS + 3\ O_2 \rightarrow ZnO + 2\ SO_2$

Balance Chemical Equation in 4 Terms

9) NaOH + H_2CO_3 → Na_2CO_3 + H_2O

10) CH_4 + O_2 → CO_2 + H_2O

11) C_3H_8 + O_2 → CO_2 + H_2O

12) NH_3 + 7 O_2 → NO_2 + 6 H_2O

Balance Chemical Equation in 4 Terms

13) KOH + HBr → KBr + H_2O

14) Na + H_2O → NaOH + H_2

15) AlI_3 + 3 $HgCl_2$ → $AlCl_2$ + 3 HgI_2

16) Rb + $RbNO_3$ → RbO_2 + N_2

Balance Chemical Equation in 4 Terms

17) $K + MgBr_2 \rightarrow KBr + Mg$

18) $I_2O_3 + CO \rightarrow I_2 + CO_2$

19) $C_6H_6 + O_2 \rightarrow CO_2 + H_2O$

20) $NaBr + CaF_2 \rightarrow NaF + CaBr_2$

Balance Chemical Equation in 4 Terms

21) $H_2SO_4 + NaNO_2 \rightarrow HNO_2 + Na_2SO_4$

22) $FeBr_3 + H_2SO_4 \rightarrow Fe_2(SO_4)_3 + HBr$

23) $CH_4 + O_2 \rightarrow CO_2 + H_2O$

Answers

1) $2Ag + H_2S \rightarrow Ag_2S + H_2$

2) $6K + 1B_2O_3 \rightarrow 3K_2O + 2B$

3) $2HCOOH + O_2 \rightarrow 2 CO_2 + 2H_2O$

4) $2NaCl + 1F_2 \rightarrow 2NaF + 1Cl_2$

5) $CH_4 + O_2 \rightarrow CO_2 + H_2O$

6) $2Co + 3H_2O \rightarrow Co_2O_3 + 3H_2$

7) $10Na + 2NaNO_3 \; 6Na_2O + 1N_2$

8) $2ZnS + 3 O_2 \rightarrow 2ZnO + 2 SO_2$

9) $2NaOH + 1H_2CO_3 \rightarrow Na_2CO_3 + 2H_2O$

10) $1CH_4 + 2O_2 \rightarrow 1CO_2 + 2H_2O$

Answers

11) $N_2 + O_2 \rightarrow N_2O_5$

12) $1S_8 + 12O_2 \rightarrow 8SO_3$

13) $4Fe + 3O_2 \rightarrow 2Fe_2O_3$

14) $16Al + 3S_8 \rightarrow 8Al_2S_3$

15) $2HO_2 + 1O_2 \rightarrow 2H_2O_2$

16) $PtCl_4 \rightarrow Pt + 2Cl_2$

17) $6Cs + 1N_2 \rightarrow 2CsN_3$

18) $S_8 + 24F_2 \rightarrow 8SF_6$

19) $4Al + 3O_2 \rightarrow 2Al_2O_3$

20) $P_4O_{10} + 6H_2O \rightarrow 4H_3PO_4$

Answers

21) $H_2SO_4 + NaNO_2 \rightarrow HNO_2 + NaSO_4$

22) $2FeBr_3 + 3H_2SO_4 \rightarrow 1Fe(SO_4) + 6HBr$

23) $1CH_4 + 2O_2 \rightarrow 1CO_2 + 2H_2O$

Balance Chemical Equation in 5 or more Terms

1) $Ca_3(PO_4)_2 + SiO_2 + C \rightarrow CaSiO_3 + CO + P$

2) $3\ Pt + HNO_3 + 18\ HCl \rightarrow H_2PtCl_6 + 4\ NO + 8\ H_2O$

3) $HCl + CaCO_3 \rightarrow CaCl_2 + H_2O + CO_2$

4) $HCl + MnO_2 \rightarrow MnCl_2 + Cl_2 + H_2O$

Balance Chemical Equation in 5 or more Terms

5) $C_5H_8O_2 + NaH + HCl \rightarrow C_5H_{12}O_2 + NaCl$

6) $S + HNO_3 \rightarrow H_2SO_4 + NO_2 + H_2O$

7) $NH_3 + CuO \rightarrow Cu + N_2 + H_2O$

8) $Cu + HNO_3 \rightarrow Cu(NO_3)_2 + NO + H_2O$

Balance Chemical Equation in 5 or more Terms

9) $HNO_3 + NaHCO_3 \rightarrow NaNO_3 + H_2O + CO_2$

10) $6\,TeCl_2 + H_2O \rightarrow TeO_2 + Te + 2\,HTeCl_6$

11) $Na_2CO_3 + HCl \rightarrow NaCl + H_2O + CO$

12) $H_3SO_4 + HI \rightarrow H_2S + I_2 + H_2O$

Answers

1) $1Ca_3(PO_4)_2 + 3SiO_2 + 5C \rightarrow 3CaSiO_3 + 5CO + 2P$

2) $3 Pt + 4HNO_3 + 18 HCl_2 \rightarrow 3HPtCl_6 + 4 NO + 8 H_2O$

3) $2HCl + 1CaCO_3 \rightarrow 1CaCl_2 + 1H_2O + 1CO_2$

4) $4HCl + MnO_2 \rightarrow MnCl_2 + Cl_2 + 2H_2O$

5) $1C_5H_8O_2 + 2NaH + 2HCl \rightarrow 1C_5H_{12}O_2 + 2NaCl$

6) $S + 6 HNO_3 \rightarrow H_2SO_4 + 6 NO_2 + 2 H_2O$

7) $2NH_3 + 3CuO \rightarrow 3Cu + N_2 + 3H_2O$

8) $3Cu + 8HNO_3 \rightarrow 3Cu(NO_3)_2 + 2NO + 4H_2O$

9) $1HNO_3 + 1NaHCO_3 \rightarrow 1NaNO_3 + 1H_2O + 1CO_2$

10) $6 TeCl_2 + H_2O \rightarrow TeO_2 + Te + 2H_2TeCl_6$

11) $Na_2CO_3 + 2HCl \rightarrow 2NaCl + H_2O + CO_2$

12) $H_2SO_3 + 8HI_4 \rightarrow H_2S + 4I_2 + 4H_2O$

Balance Chemical Equation in Word Probems

1) Aluminum metal reacts with copper chloride to produce aluminum chloride and copper metal.

2) Zinc and lead nitrate react to form zinc nitrate and lead.

3) The combustion of propane gas with gaseous oxygen gives carbon dioxide and water vapour.

4) carbon + oxygen → carbon dioxide

Balance Chemical Equation in Word Probems

5) Lead nitrate reacts with sodium bromide to produce lead bromide and sodium nitrate

6) Aluminum bromide and chlorine gas react to form aluminum chloride and bromine gas.

7) Iron oxide can be reduced by carbon monoxide gas to form iron metal and carbon dioxide gas.

8) potassium hydroxide → potassium oxide + water

Balance Chemical Equation in Word Probems

9) copper + silver nitrate → copper nitrate + silver

10) Zinc metal reacts with oxygen gas to produce zinc oxide

11) Potassium metal and chlorine gas combine to form potassium chloride.

12) ethane + oxygen → carbon dioxide + water

Balance Chemical Equation in Word Probems

13) zinc + copper sulfate → zinc sulfate + copper

14) Aluminum sulfate reacts with barium iodide to produce aluminum iodide and barium sulfate

15) Aluminum and hydrochloric acid react to form aluminum chloride and hydrogen gas.

16) Aluminum oxide reacts with hydrofluoric acid to give aluminum trifluoride and water.

Balance Chemical Equation in Word Probems

17) Sodium metal reacts with water to produce sodium hydroxide and hydrogen gas.

18) Copper and sulfuric acid react to form copper sulfate and water and sulfur dioxide.

19) Methane, ammonia, and oxygen gas react together to form hydrogen cyanide and water.

20) Aluminum oxide → Aluminum + Oxygen

Balance Chemical Equation in Word Probems

21) water → hydrogen + oxygen

22) Copper and sulfuric acid react to form copper sulfate and water and sulfur dioxide.

23) Methane, ammonia, and oxygen gas react together to form hydrogen cyanide and water.

24) Aluminum oxide → Aluminum + Oxygen

Balance Chemical Equation in Word Probems

25) Lead sulfide reacts with oxygen gas to produce lead oxide and sulfur dioxide.

26) Hydrogen gas and nitrogen monoxide react to form water and nitrogen gas.

27) barium chlorate → barium chloride + oxygen

28) sodium chlorate → sodium chloride + oxygen

Balance Chemical Equation in Word Probems

29) sodium nitrate → sodium nitrite + oxygen

30) nitrogen + hydrogen → ammonia

Answers

1) $2\,Al + 3\,CuCl_2 \rightarrow 2\,AlCl_3 + 3\,Cu$

2) $Zn + Pb(NO_3)_2 \rightarrow Zn(NO_3)_2 + Pb$

3) $C_3H_8 + 5\,O_2 \rightarrow 3\,CO_2 + 4\,H_2O$

4) $C + O_2 \rightarrow CO_2$

5) $Zn + Pb(NO_3)_2 \rightarrow Zn(NO_3)_2 + Pb$

6) $2\,AlBr_3 + 3\,Cl_2 \rightarrow 2\,AlCl_3 + 3\,Br_2$

7) $Fe_2O_3 + 3\,CO \rightarrow 2\,Fe + 3\,CO_2$

8) $2\,KOH \rightarrow K_2O + H_2O$

9) $Cu + 2\,AgNO_3 \rightarrow Cu(NO_3)_2 + 2\,Ag$

10) $2\,Zn + O_2 \rightarrow 2\,ZnO$

Answers

11) $2K + Cl_2 \rightarrow 2KCl$

12) $2CH_6 + 7O_2 \rightarrow 4CO_2 + 6H_2O$

13) $Zn + CuSO_4 \rightarrow ZnSO_4 + Cu$

14) $Al_2(SO_4)_3 + 3BaI_2 \rightarrow 2AlI_3 + 3BaSO_4$

15) $2Al + 6HCl \rightarrow 3H_2 + 2AlCl_3$

16) $Al_2O_3 + 6HF \rightarrow 2AlF_3 + 3H_2O$

17) $2Na + 2H_2O \rightarrow 2NaOH + H_2$

18) $Cu + 2H_2SO_4 \rightarrow CuSO_4 + 2H_2O + SO$

19) $2CH_4 + 2NH_3 + 3O_2 \rightarrow 2HCN + 6H_2O$

20) $2Al_2O_3 \rightarrow 4Al + 3O_2$

21) $2Na + 2H_2O \rightarrow 2NaOH + H_2$

22) $Cu + 2H_2SO_4 \rightarrow CuSO_4 + 2H_2O + SO_2$

Answers

23) $2 CH_4 + 2 NH_3 + 3 O_2 \rightarrow 2 HCN + 6 H_2O$

24) $2 Al_2O_3 \rightarrow 4 Al + 3 O_2$

25) $PbS_2 + 3 O_2 \rightarrow PbO_2 + 2 SO_2$

26) $2 H_2 + 2 NO \rightarrow 2 H_2O + N_2$

27) $Ba(ClO_3)_2 \rightarrow BaCl_2 + 3 O_2$

28) $2 NaClO_3 \rightarrow 2 NaCl + 3 O_2$

29) $2 NaNO_3 \rightarrow 2 NaNO_2 + O_2$

30) $N_2 + 3 H_2 \rightarrow 2 NH_3$

www.ingramcontent.com/pod-product-compliance
Lightning Source LLC
Chambersburg PA
CBHW080438220526
45465CB00009B/3336